The Illustrated Guide to Understanding Astrophysics and the Universe

By Charles River Editors

The Hubble Ultra-Deep Field. The equivalent area of sky that the picture occupies is shown as a red box in the lower left corner.

About Charles River Editors

Charles River Editors was founded by Harvard and MIT alumni to provide superior editing and original writing services, with the expertise to create digital content for publishers across a vast range of subject matter. In addition to providing original digital content for third party publishers, Charles River Editors republishes civilization's greatest literary works, bringing them to a new generation via ebooks.

Visit charlesrivereditors.com for more information.

Introduction

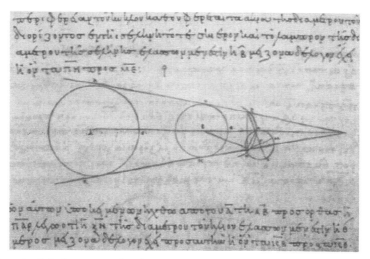

Aristarchus's 3rd century B.C. calculations on the relative sizes of the Sun, Earth and Moon

"My goal is simple. It is a complete understanding of the universe, why it is as it is and why it exists at all." - Stephen Hawking

From the dawn of time, man has sought to understand the Universe and his place in it. How did the Earth and the Solar System come to be? How was the Universe created?

Like other scientific disciplines, astronomy and astrophysics is one big detective story. Hypotheses are formed, observations taken, and experiments performed in the search for universal laws that describe all that we see. A good hypothesis or theory will make predictions of future observations, the results of which will either refute the theory, or be consistent with it. Astronomy is at a distinct disadvantage over other branches of science in one crucial way: for the most part, our observations only consist of photons (i.e. light) from far away sources, rarely can we touch and manipulate the things we observe, and thus create our own controls for an "experiment". We must wait for those far-away objects to cooperate. The light must be analyzed in many different ways (variations in space, time, intensity and frequency to name just a few), comparing different objects with one another, and making informed opinions upon the results. The light over the whole electromagnetic spectrum from a particular "target" must be explained in a consistent way using the laws of physics, and often it's back to the telescope for a new set of observations when some part of the theory proves inadequate. Or, back to some intensive computations. Nevertheless, astronomers and astrophysicists have done remarkably well over the last couple of centuries, allowing us to present an overview of how the Universe functions.

In this resourceful guide, common questions about the Universe will be explained in comprehensive but easy to understand terms. You'll learn the answers to some of the most important questions, including:

*How do stars form?
*What happens when stars die?
*What do we know about the origin of the universe?
*What is dark matter and why do we suppose it exists?
*How does our solar system fit into the Milky Way Galaxy?
*What galaxies are around us, and how are galaxies classified?
*What is the cosmological principle?

The Illustrated Guide to Understanding Astrophysics and the Universe gives an entertaining and educational overview of our Universe, from the smallest matter to massive black holes, and everything in between. Whether you are an experienced amateur or a complete novice, let The Illustrated Guide to Understanding Astrophysics and the Universe be your guide to the stars.

The Sun: The Enabler of Life on Earth

The Sun – a ball of gas around which the Earth orbits, is the source of energy that enables life on Earth, and so we have a vested interest in knowing how it works. Is this energy source reliable? Does the output of the Sun vary with time, and if so, on what times scales? These are fundamental questions to which we must know the answer if we are to have any idea what future of the Earth and solar system will be like. Remarkably, it is only within the last century that science has been able to answer these questions.

In the 1860s Lord Kelvin wrote in "On the Age of the Sun's Heat" that the Sun was powered by the collapse from its initial formation. Imagine the Sun initially as a very large, extended ball of material, which then collapses upon itself from the effects of its own gravity. We know from every day experience that objects that fall from a height gain kinetic energy (the energy of movement, $E = \frac{1}{2} mv^2$ (*m* is mass, *v* is velocity), by classical physics) and that energy can then be used for other purposes. This process is at work in a hydroelectric dam: water falls from the upper levels through a turbine, the water then pushes the blades of the turbine to generate electricity. Unfortunately, Kelvin's method predicts a solar lifetime of something like 20 million years, and the Sun is much older than that! Enter Sir Arthur Eddington who in 1920 postulated (without knowing the details) that perhaps hydrogen is fused into helium, producing a surplus of energy, which can then be radiated away as light energy. How does this work? All mass has an energy equivalent, as given by Albert Einstein's famous equation, $E=mc^2$, a consequence of his special theory of relativity, and the fact that objects cannot move faster than the speed of light. Here, $c=3 \times 10^8$ m/s is the speed of light in a vacuum, and *m* is the mass of the object in question. Multiply them together in the manner shown, and you obtain the energy equivalent for a given mass. Four hydrogen atoms are fused into a single helium atom but the resultant helium atom has less mass that the original hydrogen atoms! That extra mass-as-energy must go somewhere (total energy is conserved, a consequence of the first law of thermodynamics) and that somewhere is the light we see coming from the Sun. This turns out to be the heart of the true, complex situation, the basics of which were worked out over the next several decades by such prominent scientists as Subrahmanyan Chandrasekhar (Nobel Prize, 1983) and Hans Bethe (Nobel Prize, 1967). Hydrogen fusion will power the Sun in the current phase of its lifetime for around 10 Gyr (1 Gyr = 1 gigayear = 1 billion years). There are numerous ways that four protons can be fused to a helium nucleus. Each way uses a series of reactions, but in the Sun the dominant chain of reactions gives the net result:

$$6\ ^1H = {}^4He + 2\ ^1H + 2e^+ + 2\nu + 2\ \text{photons}$$

In other words, two protons are used as a catalyst, while the other four get "used up" in the process. The "ν" indicates a neutrino, a strange particle that has such low interaction rates with other matter that the vast majority of them will freely escape the Sun without any interference. The two photons on the other hand provide some of the "lost" mass energy, interacting with matter to produce heat. Also, the two positrons, "e+", will mutually annihilate with their antiparticle (i.e. electrons), providing more such energy.

   The current age of the Sun (and Earth) is about 4.6 Gyr, thus we don't have to worry about the Sun running out of fuel any time in the near future! Within the astronomical community, fusion is also known as *burning*, not to be confused with the chemical process otherwise known as burning.

   At about 150 million kilometers' distance, the Sun is the closest example of a star to the Earth, and serves as a prototype to which we compare all other stars, with the fundamental parameters of other stars given in solar units, e.g. solar masses ($M_\odot$) or solar luminosities ($L_\odot$). The basic physical characteristics of the sun are hard to imagine: Mass M = 1.99×10^{30} kg, total energy output $L_\odot = 3.8 \times 10^{26}$ W (or 3.8×10^{24} 100 W light bulbs!), radius $R_\odot = 7.0 \times 10^5$ km, surface gravity = 28 times that of the Earth. Even more remarkable is the variation in other quantities, such as temperature, density, and pressure, from the surface to the center of the Sun. The Sun is layered, and in *hydrostatic equilibrium* (i.e. the layers generally remain stationary): for the layers to remain stationary (most are), the downward force due to gravity from all layers above and including the layer must be cancelled out by the upward pressure from the layers below. For this condition to remain true at all depths, certain things must hold true: as depth from the surface increases, pressure, temperature and density increase dramatically. At the surface, temperature is 5800 K, density is 2×10^{-7} g/cm^3. At the core, temperature is 15 million Kelvin, density 160 g/cm^3![1]

   The various layers of the Sun from the center, outwards, are as follows: the core, comprising about 10% of the mass, in which the thermonuclear reactions are taking place; a radiative zone, in which the energy is carried outward from the core via radiation; a convective zone, in which the energy is carried outward mainly by convection; the photosphere, so called because there is where most of the photons that we see originate, and is generally considered the "surface" of the Sun; the atmosphere, subdivided into the chromosphere and corona. There are many estimates of how long the energy takes to escape the Sun, ranging from tens of thousands of years to millions of years. It is not instantaneous! The sun could shut off nuclear production (not that it would), and still shine for thousands of years before we would begin to feel the effects on the Earth. How do we know the internal structure of the Sun, given that we can only really see the surface? The answer is through complex numerical computer models that include the relevant physics. These models make predictions about what we should see, and we then compare the predictions to the observations. The models are then tweaked until they are consistent with observations. We've known the basics of stellar structure and evolution since the 1950s and 1960s from the very earliest computer models, but we only have gotten the details right in the 1990s, the first

---

1   For reference, at standard temperature and pressure, the density of water is 1 g/cm^3.

time at which science was able to produce a stellar model that simultaneously predicted the luminosity and radius of the Sun to reasonable precision. There are other things that we have yet to fully understand, such as the magnetic fields and sunspot cycles of the Sun; they are works in progress.

At a distance of 150 million kilometers (the Earth-Sun distance), the Sun delivers 1.4 thousand W/m^2 to any surface perpendicular to the Sun's rays. It is this energy that heats up the Earth's atmosphere, induces weather patterns, and drives photosynthesis for food production in plants. Energy is delivered to a surface from the Sun, d, at a rate proportional to $1/d^2$. Thus the planets that orbit the Sun interior to the Earth's orbit (Mercury, Venus) receive greater irradiance than the Earth, and those further away (Mars, Jupiter, etc.) receive lesser irradiance. The surface temperatures of the planets vary accordingly.

The composition of the Sun is about 73% by mass of hydrogen, helium 25%, and 2% all others, and reflects the composition of the cloud from which the Sun was initially formed. Later, when we talk about stellar evolution and cosmology, you will learn that this means the Sun (and the rest of the solar system) was formed from material from earlier-generation stars, as that 2% of "other material" can only be formed through the processes of stellar evolution, and did not originate with the Big Bang. And a side-note of caution: you will often hear astronomers talk about "metals" or "metallicity". These terms refer to all material that is *not* hydrogen and helium (the two lightest elements), and has nothing to do otherwise with the electrical properties of the elements in question, and typically makes everybody who encounters it wince the first time.

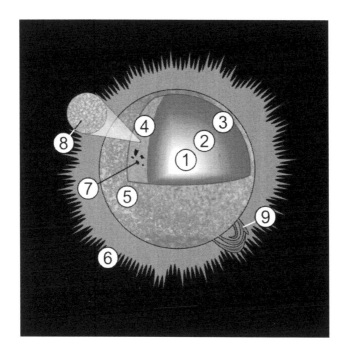

Illustration 1: Structure of the Sun: 1) Core, 2) Radiative zone 3) Convective zone 4) Photosphere 5) Chromosphere 6) Corona (hot, tenuous outer later, only visible from Earth during a solar eclipse), 7) Sunspot (transitory, dark spots caused by twisted magnetic field lines) 8) Granules (caused by surfacing of convective cells) 9) Prominence (transitory, cause of solar storms on Earth).

The output from the Sun does vary a bit throughout the sunspot cycle of 11 years (the magnetic field of the Sun reverses polarity every 11 years), and perhaps on longer cycles as well; how this affects Earthly weather and climate is still a work in progress, so stay tuned! The baseline of good measurements over time (using satellites) is too short to have gained a firm handle on any potential effects.

As for long-term variability (millions to billions of years), that is best left for the sections on star formation and stellar evolution!

The Hertzsprung-Russell diagram

Along with dust and gas, stars are one of the basic building blocks of the galaxy, with the stellar population providing most of the visible light in the universe.

It is useful to be able to characterize stars in various ways to pick out similarities and differences. Developed by Ejnar Hertzprung and Henry Norris Russell in the early 1900s, the *Hertzprung-Russell diagram* (H-R diagram or HRD) is a useful way of characterizing stars, a graph that plots the intrinsic luminosity of the star versus its surface temperature. Warning: for historic reasons, the temperature scale is *reversed*, with temperature increasing to the left![2] Alternatively, the x-axis might label stellar spectral class, a way to classify stars based their observed light spectra (spectra and temperature are intricately linked). The y-axis may also alternatively label "absolute magnitude", a reverse logarithmic system that is equivalent to luminosity, developed in the early days of modern astronomy. Regardless of the strangeness of the labelling system, when stars are plotted on the HRD, groups of stars immediately become obvious, and those different groups of stars generally label stars that are in a different phase of their evolution. The most prominent feature is the Main Sequence (MS), an approximately diagonal band across the HRD that are stars that are undergoing hydrogen burning in the center of the star. Within the MS, a star's position is determined via its mass: more massive stars burn hotter and brighter, and so are higher up and to the left. Determining a star's position in the HRD is fraught with difficulty, sometimes creating considerable uncertainties, and problems with interpretation. Nevertheless, the HRD is one of the most basic diagrams regularly used by astronomers, and the wide range of stellar characteristics is immediately obvious from the use of logarithmic scales on both axes.

---

2 One of the fastest ways to confuse an astronomer is to reverse the x-axis on one of the graphs so that it is "normal"!

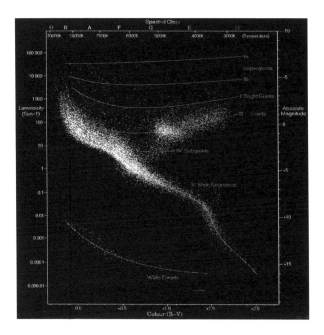

Illustration 2: Hertzsprung-Russell diagram illustrating how different classes of stars are related via luminosity, colour, spectral class, and absolute magnitude

Star Formation

One of the most fundamental questions we can have is: How did the solar system come about? Developing a coherent theory for star formation has been one of the most challenging problems in all of astronomy. Part of the problem is that stellar births happen in dense environments, so dense that photons originating from the incubating stars cannot penetrate the surrounding environment to reach our telescopes. The light gets absorbed and reemitted at longer wavelengths (i.e. At lower energies), and we only infer the new stars' existence indirectly. Stars form from very dense clouds (compared to the rest of interstellar space) of *dust* and *gas*. (Here "dust" refers to material made up of more than one molecule.) In order for a star to form, the self-gravity of a cloud of material must be able to overcome any internal pressures the cloud might have. In general, the material must cool sufficiently to reduce the pressure, so that gravity can then take over. If a region has too low a density, external sources of heat (e.g. photons from outside the cloud) will counteract any cooling and prevent star formation. Only within the disk of the Milky Way galaxy are cloud densities and sizes large enough to be self-shielding from external sources of heat, and high enough to support star formation. Even here, usually cloud densities only become large enough for star formation through some external mechanism – a shock wave might travel through the cloud, or two less-dense clouds might collide to form a higher-density cloud that can then form stars. The impressive images of nebulae generally depict regions of active star formation.

Illustration 3: Composite three-color image of the M 17 star-formation region taken with the ISAAC near-infrared instrument at the 8.2-m VLT ANTU telescope at Paranal. The darkest regions would hide still forming stars. The glow is provided by the light of the largest, brightest stars.

Once a collapse does happen, a small core initially forms (a fraction of a solar mass is our best guess, maybe 0.01 to 0.001 $M_\odot$), upon which any following material accretes. Our best guess is that accretion stops when either the birth cloud is depleted, or some other event disrupts the cloud. At this point, the star emerges from its birth cloud, and we should be able to see it. The result is an object ranging from a brown dwarf (an object not massive enough to fuse hydrogen to helium at its core), to the least massive stars (0.08 $M_\odot$) to stars typically up to 20 $M_\odot$. More massive stars do exist (100 $M_\odot$?), but their formation mechanism is even more of a mystery to us – the most massive might be the product of stellar mergers. Stars form in a hierarchy of masses, called the *initial-mass function* (IMF), in which smaller-mass stars vastly outnumber higher-mass stars. The individual birth clouds of stars are sub-clouds of a larger *star-formation region* (SFR) in which multiple stars tend to be formed together in *open clusters*. Once the largest stars are formed, few in number that they are compared to the other stars, their radiation tends to dominate the region (they are extremely bright and hot), heating up and dissipating the clouds, preventing further star formation in the immediate region. The largest stars have a very short lifespan (a few million years), upon which time they *supernova* (explode), sending out shock waves into surrounding regions, perhaps triggering further episodes of star formation.

Within an individual star system, during final phases of formation, dramatic events take place. One of the principles that works against the formation of a star is that of the conservation of angular momentum. Any extended object (the initial gas cloud) that tries to compress down to something as small (relatively speaking) as a star will have sufficient angular momentum (spin) to prevent the collapse from happening – the star won't be able to hold itself together! The accreting material must get rid of its angular momentum somehow. This is thought to happen through magnetic fields accelerating a fraction of the material out along the spin poles of the system in the form of collimated *jets*. Jets are observed in many SFRs, we think they carry the access angular momentum away from the system, and allow accretion to continue. The accreting material forms a *disk* around the star, and in the final phase of accretion, a remnant debris disk is left over, from which a planetary system may be formed. Further complicating the picture is that binary (or higher multiple) star systems may be formed that will influence the formation of any planetary system. Our own solar system contains a single star, whereas our closest stellar neighbor, Alpha Centauri, is actually a trinary system, with two stars (Alpha Cen A and B) orbiting close to one another, and a third (Proxima Centauri) orbiting the pair from a distance. So, SFRs are highly volatile, dangerous places!

The Basics of Stellar Evolution

The evolutionary future of a single star (i.e. not in a close binary) is dependant mainly on the initial mass of the star. Upon emerging from their birth cloud, stars with mass less than about 5 $M_\odot$ have what is known as the *pre-main-sequence* (PMS) phase of evolution, they have yet to start core hydrogen burning. Yet, they are actually quite bright, so what is making them shine? Lord Kelvin had the right idea: initially these stars shine via the conversion of gravitational potential energy into heat, which then gets radiated away as light. As the stars slowly shrink over the period of several million years, eventually the core become dense and hot enough to commence thermonuclear reactions, the shrinking stops, and the star reaches the main sequence of core hydrogen burning. For stars with mass greater than about 5 $M_\odot$, core fusion begins before the main accretion phase has finished, and so we never see the star in a PMS phase, it appears immediately on the main sequence when it emerges from its birth cloud.

The MS is the longest phase of a star's lifetime, but the length of the phase varies dramatically according to the mass of the star. A star's luminosity on the MS varies approximately as $M^{3.5}$, but the amount of fuel available to the star varies approximately linearly, so more massive stars consume their fuel at a much higher rate than less massive stars, resulting in shorter MS lifetimes. The MS lifetime varies approximately as $M^{2.5}$. The lowest-mass stars have lifetimes that are trillions of years long, much older that the age of the universe, while solar-mass stars have MS lifetimes around 10 Gyr, and the highest-mass stars only have lifetimes of a few million years. One consequence of the latter, is that if we see any bright, massive stars in a region, we know that star formation took in the recent past. Indeed, the life time disparities between low- and high-mass star is so high that a high-mass star formed at the same time as a low-mass star can go through its entire evolutionary cycle before the low-mass star even reaches the main sequence!

What signifies the end of the MS? The core of the star (the only place hot and dense enough for hydrogen to burn) runs out of hydrogen, and as the residual energy drains out of the core, the core begins to shrink. The region *around* the core follows the core inward, and becomes dense enough to start hydrogen burning, a phase known a *shell-hydrogen burning*. Strangely enough, this shell phase has a *higher* energy production rate than the MS phase. The higher luminosity of the star causes the outer layers to expand, the surface appears cooler, but overall the star is brighter – the star has entered the *red giant* phase of stellar evolution (see HRD).

Stars as Black Body Radiators

Stars radiate energy as approximately a *black body*, a term with a very specific meaning in physics. The energy per unit surface area per unit time (energy *flux*) of a black body of temperature T is given by $F=\sigma T^4$ where σ is the Stefan-Boltzmann constant. If we assume the star is a sphere with radius r, and surface area $A=4\pi r^2$, then the total luminosity of a star is $L=FA=4\pi r^2\sigma T^4$. A small, hot star can be just as luminous as a very large but cool star. The trade-off between surface temperature, radius, and total luminosity often confuses people, as we normally associate hotter objects as brighter. This is true for a *patch* of the surface, but integrated over its surface a large, cool object might have more power output than a small, hot object. A star with a surface temperature greater than the Sun's (5800 K) looks blue, a star that is cooler appears red, hence the loose terminology *blue* meaning *hot*, and *red* meaning *cool*. Also, *bluer* means *hotter*, *redder* means *cooler*.

The red-giant phase for a solar-mass star is very bad news for any planetary system the star might have. Not only is the star much brighter, but it is also much larger, with a radius equivalent to that of the order of Jupiter! Planets will get baked, and any inner planets will get absorbed by the host star. So, while the core of the star has increased in density, overall the star is much less dense than it was before, with only a tenuous gravitational hold on its outer layers. Stars with a mass less than 0.5 $M_\odot$ will basically fade away at this point as helium white dwarf stars. Stars with a mass between 0.5 And 1.8 $M_\odot$ undergo a process known as *helium flash*, a very short-lived (minutes!), dramatic event whereby suddenly the conditions are ripe to (somewhat explosively) start burning helium into carbon in the core. The energy output spikes and radiation pressure increases enough to throw off a large part of the outer layers of the star (remember that tenuous hold?), before settling down to a steady-state core-helium-burning phase of evolution.

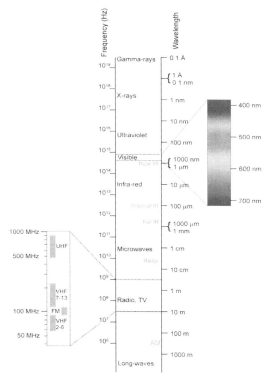

Illustration 4: Electromagetic spectrum

Stars with an initial mass greater than 1.8 $M_\odot$ don't have it quite so dramatically and will start burning helium smoothly. Regardless of mass, for stars in which core helium burning takes place, shell hydrogen burning may still also occur. After about 1 Gyr, helium in the core runs out, and (once again) the core shrinks, initiating concentric shell burning of both helium and hydrogen with the latter feeding the former. For stars with initial mass less than 10 $M_\odot$ they will become brighter, and will enter the *asymptotic giant branch* (AGB) of evolution, almost indistinguishable in the HRD from the red-giant branch. For these stars their cores will not become dense enough to fuse the carbon and oxygen, of which it is now composed, into heavier products, instead the last bits of shell hydrogen and helium will be burned with such intensity that any remaining outer layers of material will be thrown off in the form of a *planetary nebula*, so called because on low-resolution images they resemble nebulous planets like Uranus and Neptune (they were named by early astronomers before they knew what they were seeing). What was once the core of the star is now become the surface! Moreover, what was once an approximately 10 $M_\odot$ star has been reduced to a maximum mass of 1.4 $M_\odot$, and 8^+ $M_\odot$ has been returned to the interstellar medium (ISM). Such stars will become carbon-oxygen (CO) white dwarfs, with a radius similar to that of the Earth, and typical masses of 0.5 to 1.4 $M_\odot$ That will slowly cool over time to dim object. The material that has been ejected from the star is returned to the ISM, much of it enriched with the products of stellar evolution, and will probably find itself incorporated into later generations of stars.

Illustration 5: The Ring planetary nebula in infrared from NASA's Spitzer space telescope. Through an optical telescope at low resolution it would appear similar to the planet Uranus.

For stars larger than 8 to 10 $M_\odot$ (the dividing line is not clear!), the fusion process continues, with carbon and oxygen fusing to neon. Further outer layers of the star may be lost. Stars that retain sufficient mass will subsequently fuse neon to magnesium. Eventually silicon will be formed, which will then be fused to the Iron-56 (^{56}Fe), the isotope with the largest-known binding energy. Up to the production of ^{56}Fe, all fusion processes *liberated* energy, and so it was energetically favorable for the process to take place. However, it is not energetically favorable to produce elements heavier then ^{56}Fe through fusion, so the process stops. Unfortunately, it was only the heat produced by fusion that was keeping the core from shrinking, so once the fusion stops, trouble! The core collapses becomes so dense and hot that protons and electrons are brought together to form a neutron, and the core becomes a neutron star, with a mass between 1.4 to 3 M_O, and a relatively rigid structure only a few kilometers in radius. This is much smaller than the core from which the neutron star originated, so the outer layers of the star infall briefly, and then "bounce back" in a *supernova* explosion. All outer layers are shed back into space, returning processed material to the ISM. The energy output of such an event is enormous, with the output of one supernova often outshining that of the host galaxy for a short period of time.

During this violent time the outer layers of the star are bombarded by neutrons, alpha particles (helium nuclei) and protons. The neutrons, in particular, have a very interesting effect, interacting with the nuclei of various elements, causing heavier and heavier elements to be formed. All of the other known elements in the universe, particularly the ones heavier then ^{56}Fe are produced in these events, even familiar elements such as gold, silver, and uranium. And so our story about stellar evolution has been leading up to this very important result: *without stellar fusion and supernova explosions, the materials that went into making the Earth, Sun, and Solar System would not exit, and life as we know it would not be possible.*

Illustration 6: Sirius A and B binary system as observed by the Hubble Space Telescope. Sirius A (the bright component) is a 2.02 M_Θ MS star, whereas Sirius B (small dot to the lower left of the Sirius A) is a 0.98 M_Θ white dwarf

Binary Star Systems

It is quite common for stars to be formed as binary systems. All ranges of separations are possible, from systems that are just a few stellar radii apart, to stars of such large separation that they are essentially two separate systems that can take centuries to orbit about their common center of mass. The mass difference between the two stars can be quite different, meaning that the two will evolve at completely different rates. Previously we chronicled the lifetime of single stars, during which the red giant phase of evolution in particular, a star expands to an enormous radius compared to its MS radius. Imagine if this star is in a binary system close enough for the star's expansion to intrude upon the territory of the other star. Where does this point occur? Imagine a line connecting the two stars. Along that line, somewhere, is a point in which the gravitational force between the two stars is equal – put an object there, give it a small perturbation, and it could fall into (or orbit) either star, depending upon direction of that perturbation. Should one of the two stars enlarge to that point, material from the growing star will fall onto the other star, disrupting the course of events for *both* stars. The actual sequence of events depends heavily upon the relative and absolute mass of the two different stars, coupled with their initial separation. The intermediate and end products constitute some of the most interesting objects within our own Galaxy. Some possibilities:

White dwarf with expanding star – the material from the growing star will land on the white dwarf, and periodically undergo nuclear burning on the surface. The burning phase is short lived and explosive, called a *nova* (as opposed to a supernova), with the luminosity of the star increasing by several orders of magnitude. The nova can be reoccurring.

Supernova explosion – if the white dwarf mass should exceed 1.44 M_Θ as a result of accretion, the white dwarf will collapse to a neutron star in one type of supernova explosion. Given the (somewhat) uniform conditions under which this particular type of supernova occurs, all such supernovae are thought to have similar luminosities, and hence are used as *standard candles* for measuring distances to the event in question. If the binary system survives, a neutron-star/expanding-star binary will result, with perhaps material now falling on to the neutron star. Like the supernovae that occur in massive stars, this type can temporarily exceed the luminosity of the host galaxy.

X-ray binary – so called because they were initially detected in x-rays. This is a small black hole (maybe a few M_Θ in size) orbited by an expanding star. The black hole might have been formed by the core collapse of a massive star, or by mass accretion on to a neutron star, which then exceeded the upper-mass limit of a neutron star of a few solar masses (the exact upper limit is not yet known), and then collapsed to a black hole. The x-rays are emitted by the in-falling material.

The Solar System

Now that we know where all of the heavier elements come from, we can move on to examining the make-up of our own solar system. There are eight major planets directly orbiting the sun, comprised of the four inner *rocky* planets, Mercury, Venus, Earth, and Mars, plus four outer *gas giants*, Jupiter, Saturn, Uranus, and Neptune. The four inner planets are relatively small and dense (3.9 to 5.5 g/cm^3 average density), the four outer planets heavy, but large, and not so dense (0.6 to 1.6 g/cm^3 average density). Until recently, Pluto was listed as one of the major planets, it is now considered a dwarf planet.[3] Pluto is one of many dwarf planets of rock and ice that have been discovered beyond the orbit of Neptune that comprise the *Kuiper belt*. Between the orbit of Mars and Jupiter is the *asteroid belt*, a group of smaller rocky objects (asteroids) that orbit the Sun on their own. Despite the large number of objects orbiting the Sun, the vast majority of the mass within the Solar System is contained within the Sun, the total mass of orbiting material is only 0.0014 $M_\odot$!

Illustration 7: Planets of the Solar System to scale.

---

3   The confusion surrounding the status of Pluto stems from the difficulty in determining its true size compared to the other planets – it was once thought to be much larger than it is!

Kepler's Laws of Planetary Motion

An object's orbit about the Sun is determined by *Newtonian gravity*, and *Newton's three laws of motion*, summed up in *Kepler's three laws of planetary motion*: 1) The orbit of each object is an *ellipse*, with the Sun located at one of the two *foci* of the ellipse; 2) A line joining the planet and the Sun sweeps out equal amounts of area per unit time regardless of the planet's position in the orbit. (If the planet is closer to the Sun, it is moving faster than if it is further away, and is reflection of the law of *conservation of angular momentum*.); 3) The square of the orbital period is proportional to the cube of the *semi-major axis* of the ellipse. So, unsurprisingly, since all of the major planets have orbits that are close to circular, the further away the planet from the Sun, the longer its orbit.

The standard measurement of distance within the Solar System is the *astronomical unit* (a.u.), the length of the semi-major axis of the Earth's orbital ellipse about the Sun, essentially the Earth/Sun distance. One a.u. equals 149.6 million kilometers. Given that the speed of light in a vacuum, c, is 3×10^5 km/s, it takes light 8.32 minutes or 499 seconds to travel from the Sun to the Earth.

For contrast, Mercury's orbit has a semi-major axis of 0.39 a.u., and an orbital period of 88 days (0.24 years), whereas Neptune's orbital stats are 30 a.u. and 165 years respectively!

Not including the Earth, the other inner six planets have been known since antiquity, as they are visible with the naked eye, however their nature as planets orbiting the Sun was not known until Galileo first pointed a telescope at them in the early 1600s. Prior to the 1600s they were merely known to be different from other stars by the fact that they moved with respect to the "fixed" stars in the night time sky. It took the advent of the telescope for the other planets to be discovered, Uranus in 1781, Neptune in 1846, and Pluto in 1930.

Unfortunately, due to limited space, it is not possible to list in detail all of the characteristics of the planets and their respective moons, each one is worthy of a separate article all on their own! Instead, a broad overview will highlight the physics that goes into shaping the solar system.

Natural Satellites

Aside from Mercury and Venus, all of the major planets, and even the some of the dwarf planets are known to have natural satellites (moons) of some sort. Of course, the most famous is the Earth's own Moon, orbiting the Earth approximately once a month, and responsible for raising tides within the oceans of the Earth. The Earth-Moon system has the distinction of having the largest ratio of moon-to-planet mass of the major planets. The shape of the moons and asteroids fall into two broad categories, either irregularly shaped or spheroid objects. Irregulars are small, rocky and/or icy objects that are held together mainly by intermolecular forces. For more massive objects, self-gravity shapes the object into a spheroid, overcoming the intermolecular bonds, with hydrostatic equilibrium determining the object's internal structure.

Neptune's moon, Proteus, is the largest-known irregular object, with a mean radius of 210 km. Uranus' moon, Miranda, is the smallest spheroid moon, with a mean radius of 236 km. See the table below for the number of spheroid versus irregular moons for each of the planets with moons, current as of August 2012. The spheroid moons are large enough such that all should have been discovered by now. However, irregular moons can be quite small, and hence new ones are regularly discovered, so stay tuned! Saturn's largest moon, Titan, is the only moon with a substantial atmosphere (molecular nitrogen and methane).

	Earth	Mars	Jupiter	Saturn	Uranus	Neptune
Spheroid	1	-	4	7	5	1
Irregular	-	2	62	55	22	12

Atmospheres of Planets

The atmospheres of objects are determined by a number of different interacting factors: the surface gravity, distance from the Sun, and magnetic field of the individual planet. The distance from the Sun determines the amount of heating to which the planet will subjected, and thus how much kinetic energy gas particles will have. On average, in a gas, regardless of the mass of an individual particle, a particle will receive the same amount of random kinetic energy (heat), and so lighter particles will travel faster than heavier particles. The surface gravity then determines how fast a molecule or atom needs to be in order to escape the planet's surface and be lost from the atmosphere – the higher the surface gravity, the higher the speed needs to be. Thus, within any atmosphere it will be easier to lose a light atom (hydrogen or helium) versus a molecule (carbon dioxide or water), although if the surface gravity is large enough, not much of anything will be able to escape. A magnetic field helps to shield a planet from harmful radiation or particles that may serve to disassociate molecules into lighter atoms. Unshielded, molecules may disassociate, and the resultant atoms escape. All of these factors conspire to leave Mercury without anything but a trace atmosphere (very hot surface but low gravity), Venus with a hot, dense atmosphere of CO_2 and N_2, Mars a cold, sparse atmosphere of CO_2 and N_2, and the gas giants with atmospheres of H_2 and He. Uranus and Neptune also contain methane (CH_4). That leaves the Earth, with an atmosphere of

N_2, O_2, Ar, and about 1% H_2O vapour (plus liquid and frozen water on the surface), making it the only planet capable of sustaining life as we know it, as *liquid* water seem to be the key ingredient.

Ring Systems

The gas giants all have ring systems, the most famous is that of Saturn. The rings are generally thin systems of particles of ice water, the particles ranging in size from specks of dust to 10 m in diameter. The rings are probably transitory in nature, perhaps the result of destroyed moons (particularly in the case of Saturn's rings), or comets. Except for Saturn's rings (discovered by Galileo in 1610), the rings are not extensive enough to be easily observed. Jupiter's rings were discovered by the Voyager 1 probe in 1979, Uranus' rings by ground-based observations in 1977, and Neptune's rings by the Voyager 2 probe in 1989.

Water

As mentioned, *liquid* water seems to be required for life as we know it. *Frozen* water is fairly common in the solar system, existing in comets and dwarf planets, on the Moon, and on Mars, and on many of the large moons of the gas giants. It *is* possible that liquid water exists below the surface of some of the large moons of the gas giants, but space probes that land on the moons will probably be required to test this theory. Will that mean life? Once again, probably only space probes will be able to tell us the answer.

The Milky Way Galaxy and the Sun's Position within It

Galaxies are collections of dust, gas, and stars of differing shapes and sizes. Due to their extreme distances, their nature as separate systems, external to own region of dust and gas, was only confirmed in the 1920s! Galaxies range in size from small, dwarf galaxies of 10^7 stars to massive conglomerates of 10^{14} stars. Spiral and elliptical galaxies are the most well-known type of galaxies. Spirals are usually the most visually spectacular, showing intricate structure with multiple arms, with lots of bright dust and gas. How does this relate to the Solar System's position in the Universe? If you are lucky enough to visit a truly dark site in the countryside on a clear night, you will often see a band of milkiness across the night sky, with a particular large bulge in the direction of the constellation Sagittarius. This is the Milky Way Galaxy, the galaxy in which our own Solar System is located. The "milkiness" is simply many stars viewed from such a distance that it appears to be a continuous cloudy film upon the sky.

Illustration 8: Panoramic view of the Milky Way galaxy. The bulge in the direction of Sagittarius is in the middle.

Large amounts of meticulous research have revealed that the Milky Way is a large, spiral galaxy, and that the Solar System is embedded within one of the spiral arms, about 27000 light years from the center of the galaxy. Determining the Sun's position within the Galaxy, and the nature of the Milky Way has been a large challenge, very much a case of not being able to see the metaphorical forest through the trees! We are able to look at other galaxies in the universe, and immediately determine their structure, but from our vantage point in the Galaxy we are not able to see all aspects of it, much of our view remains blocked by nearby collections of dust and gas clouds. Indeed, there are many aspects of our own galaxy that remain a mystery to us, for example, how many spiral arms it has. Regardless, we've managed to figure out many things, for which we will give a brief overview.

The Parsec

One *light year* is the distance that light travels in one year, and is the most often quoted unit of distance in popular astronomy articles, e.g. the closest stellar system to the Solar System is the Alpha Centauri system, at about 4.5 light years' distance. However, astronomers prefer the *parsec* (pc), so called because of its definition. It is based upon the concept of *parallax*, which is the measure of the apparent change in direction (measured in units of angle) to a distant object due to a change in position of the observer. If the change in position of the observer is known, and the parallax angle measured, then the distance to the object can be calculated using simple geometry.

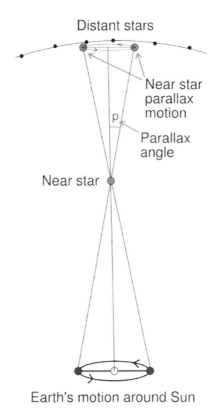

Distant stars

Near star
parallax
motion

p

Parallax
angle

Near star

Earth's motion around Sun

**Illustration 9: Parallax measurements. Far away stars or objects would remain
fixed upon the sky, whereas nearby stars would appear to move around as the Earth
orbits about the Sun.**

To measure the distance to nearby stars, we use the Earth's orbit as our base line,
measuring the change in direction to a particular star over the course of a year as we orbit
about the Sun. With a base line of one astronomical unit, an object at one **parsec** distance
will have a **par**allax of 1 *arcsecond*. (there are 60 *arcminutes* in one degree of arc, and
sixty arcseconds in one arcminute). Alpha Centauri has a parallax of 0.747 arcseconds, for
a distance of 1.34 pc. The distance of the Solar System to the center of the Galaxy is about
8300 pc, or 8.3 kiloparsecs (kpc). Unfortunately, these are hard measurements to make
accurately, and only stars within a few hundred parsecs are close enough for parallax
measurements, as the further away an object, the smaller the parallax becomes for a given

base line. For more distant objects, other, secondary, techniques must be used to determine distance.

The Galaxy has a number of principle components.

1) A disk component, about 200 pc thick, extended 20 kpc from the center, containing most of the dust, gas, and young stars of the Galaxy. The spiral arms and star-formation regions of the Galaxy are contained within the disk. Our Solar System is also located within the disk, but all of the "debris" within the disk make it hard to observe other parts of the disk unless they are relatively close by. On Earth, the equivalent situation would be a very thin, but dense layer of cloud – within the layer of cloud it is hard to see, but above and below the cloud observations become relatively easy. The disk stars orbit the center of the galaxy in close-to-circular orbits at about 250 km/s.

2) A central bulge of about 1 kpc in radius, at the very center of which is a massive black hole of about 10^6 $M_\odot$ in size. The central regions of the Milky Way block our view of the opposite side, so we don't know what opposite side of the Galaxy looks like. Several massive stars are seen to orbit the central black hole at very close range, the motions of which allow us to actually estimate the mass of the central black hole.

3) A spheroid halo of very old stars and globular clusters, extending about 50 kpc from the galactic center. These are the oldest-known stars, significantly older than the Sun, and mostly pre-dating the formation of the galactic disk. Globular clusters are star clusters containing upwards of 100 thousand to 1 million very old stars that are very useful tools in the study of stellar evolution. The individual stars and globular clusters tend toward plunging (highly elliptical) orbits around the center of the Galaxy.

4) Although not necessarily part of the Milky Way, the Galaxy is surrounded by a number of *dwarf galaxies* that are otherwise interacting with the Galaxy, and will perhaps eventually be incorporated into the Milky Way.

Galaxy formation is a topic of intense ongoing research, employing many expensive computational models. The truth is that the exact evolutionary path of the Galaxy is far less well known than the evolutionary path of any single stars we see within the Galaxy! We know that the halo and globular clusters formed first around 14 Gyr ago, but the actual appearance of the Galaxy at that point in time probably bears no resemblance to what it actually looks like now, the Universe was very young at that point in time, and organized spiral galaxies as we know them had yet to form. We know this from observations of very distance, and hence young galaxies (given the length of time the light took to reach us). At the time when the halo formed the Milky Way was probably a collection of immense, disorganized tendrils of dust and gas, furiously forming stars as whole streamers of material collapsed into a central large ball. The disk would have yet to have formed. The halo stars' orbits about the center have been largely randomized by the violent interactions that would have been occurring at the time, hence the spheroid cloud appearance the stars have collectively today. Any remaining gas (or gas that was returned to the galaxy via the processes of stellar evolution) eventually formed a disk, a natural process for clouds given that two clouds that encounter one another tend to stick together, with subsequent motion dictated by the laws of conservation of momentum and angular momentum. Conservation of angular momentum eventually leads to a thin disk rotating about the central bulge of the Galaxy.

Interestingly, halo stars have far less metals than the disk stars do, indicating that the former must come from an earlier generation of star than the latter. Confusingly, disk stars are known as Population I stars, whereas the older, halo stars are known as Population II stars. Both populations of stars can be found in the solar neighborhood, but detailed observations of the motion and composition of a star are needed before a classification can be made. Halo stars will be moving with respect to the Sun at a very high rate, whereas disk stars will share a large part of our orbital motion around the center of the Galaxy, and hence won't move very much with respect to the Sun, especially compared to the halo stars.

Galaxies around Us

The Milky Way does not inhabit the universe all by itself! The Local Group of galaxies is a group of 50+ galaxies about 3 megaparsecs (Mpc) in diameter that includes the Milky Way galaxy, and all of its nearest neighbors. The Milky Way and Andromeda (also known as M31 for Messier catalog[4] object 31), another larger spiral galaxy, dominate the Local Group. The third-largest member of the group is M33, a smaller spiral galaxy. M31 and M33 are the only large galaxies that can be seen with the naked eye in the right viewing circumstances.

The rest of the Local Group are much smaller *irregular* and *dwarf elliptical* galaxies, many of them satellite galaxies to the much larger spirals. Two of the most famous of these are two satellite galaxies of the Milky Way, the Small and Large Magellanic, easily visible from the Earth. Many of these smaller galaxies will interact and eventually merge with the larger galaxies.

Andromeda and the Milky Way appear to be moving towards one another, with a collision expected in about 4 Gyr. No two stars are expected to collide, however tidal and gas interactions will mean significant disruptions to both galaxies, with the end a result a possible merger of the two galaxies.

---

4   A list of 110 objects compiled by Charles Messier, originally published in 1771 (with subsequent revisions), "Catalogue des Nébuleuses et des Amas d'Étoiles", or "Catalogue of Nebulae and Star Clusters". It contained a list of open clusters, globular clusters, external galaxies, and Milky-Way nebulae.

Galaxy Classification

Galaxies are classified by morphology. The main classes are: *spiral, elliptical*, and *dwarf galaxies*. Spirals are the most impressive looking, with spiral arms of lots of dust and gas illuminated by bright, newly formed, massive stars, and are very blue in color. Spirals are sub-classified by: whether a central bar exists, or not; how tightly would the spiral arms are about one another; the size of a central bulge with respect to the spiral arms. The spiral arms form a very flattened disk, with the disk rotating about the center of the galaxy. Elliptical galaxies are...ellipses! They may be spherical-to-fairly-flattened in shape (note that a flattened ellipsoid may appear circular if viewed from a certain direction), with the stars orbiting the center of the galaxy from many random, highly-elliptical orbits. Ellipticals tend not to have much dust and gas, and very little, if any, active star formation. They tend to be redder in color than spirals, for as stellar populations age, red stars dominate. Ellipticals probably formed through a collision process that randomized the stellar populations of the parent objects without leaving sufficient dust and gas around to form a spiral. Dwarf galaxies may be elliptical or irregular in shape, the irregulars often having dust and gas lanes with some ongoing star formation. By number, dwarf galaxies are by far the most numerous.

Galaxies that closely interacting with one another may defy the above classification scheme, as they will often be in some intermediate state between one type of galaxy and another.

Illustration 10: A non-barred face-on spiral galaxy, M101, also known as the Pinwheel

Galaxy.

Illustration 11: A giant elliptical galaxy: Centaurus A (with dust lane, probably left over from a galaxy merger).

Illustration 12: Dwarf Irregular galaxy: The Large Magellanic Cloud, one of the Milky Way's satellite galaxies.

Galaxy Clusters, Superclusters and Filamentary Structure.

On a larger scale, clusters of galaxy with 1000+ members are not uncommon. The Virgo Cluster is the closest such cluster a cluster of 1100+ members of about 4 Mpc in diameter, and about 50 Mpc away. As we zoom out further and further, we find the Local Group is linked to the Virgo Cluster, and both are members of the Virgo Supercluster. In turn, superclusters are linked to one other in an immense filamentary structure that define the overall, current, structure of the universe. The universe, however, has had a finite lifetime, and it did not start out that way. Ongoing research involving lots of sophisticated observations, computer simulations, and theoretical calculations have unravelled many of the details, but there is still a long way to go.

The Expansion of the Universe and the Hubble Constant

One of the most astonishing discoveries of the 20th century was that of the expansion of the universe by Edwin Hubble in 1929, confirmation of a prediction made by Georges Lemaître in 1927 using General Relativity. Hubble simultaneously measured the distance and relative velocity of several distant galaxies, discovering that more distant a galaxy is, the faster it is moving away from us, implying continuous expansion. The currently accepted value for the expansion of the universe (known as the Hubble constant) is around 72 km/s/Mpc. That is, for every megaparsec a galaxy is more distant from us, it then appears to be moving away from us at a speed 72 km/s faster than it would be otherwise. This discovery implies many things regarding the history of the Universe. Firstly, it implies a distinct beginning, much more crowded and hot than the current condition of the Universe! Secondly, those initial hot conditions imply that the universe has evolved into a complex object from much simpler beginnings, the details of which we are only beginning to unravel. Thirdly, the future of the Universe depends upon such things as the density and expansion rate of the Universe, with density being particularly hard to determine.

Red/Blue and Doppler Shift

You will often hear an astronomer refer to the *redshift* of an object, a reference to the measured Doppler shift of an object that is moving away from the Earth. Doppler shift is the change in frequency of a periodic signal (could be a sound wave, or in this case, electromagnetic radiation) that is detected by an observer because the source is moving with respect to the observer. The phenomenon is heard in everyday life on a city street – listen to the motor of an approaching car, the frequency of the car, as detected, is *higher* than if the car was stationary. After the car is past (and moving away from the observer), the sound has a *lower* frequency than that of the stationary vehicle. In the visible part of the electromagnetic spectrum, blue light has a higher frequency than red light, hence an object moving away from us has a detected spectrum moved towards the *red* part of the spectrum, and we generically give this the term *redshift*. Occasionally, *blueshift* will be used to indicate the velocity of an object moving towards us, as in this case the shift will be towards the *blue* end of the spectrum. Given the expansion of the universe, only relatively nearby objects, like stars in our own galaxy, or select members of the Local Group of galaxies exhibit blueshifts, so the term is far less common then redshift. Redshift is often indicated by the letter "z". Doppler shift will only measures the radial component of a velocity with respect to the line of sight, not any transverse motion the object might have. Redshifts are measured from detailed electromagnetic spectra taken of astrophysical objects. Emission or absorption lines of known chemical species with known rest-frame frequencies are observed within the spectra of an object at some displaced frequency. By comparing the known and displaced frequency, the relative velocity or redshift can be easily calculated.

A larger Hubble constant implies a younger universe – working backwards in time, the higher the Hubble constant, the faster you collapse back to the singularity from which everything emanated, the time of the so-called Big Bang! As recently as the 1990s a wide range of Hubble constants were measured by various groups of researchers, ranging from 50 to 100 km/s/Mpc. The high values were particularly problematic, implying an age for the Universe that was younger than the oldest known stars! Fortunately, the sources of error were discovered, both in the measurements of the Hubble constant, and in the stellar-evolution models, and so all known stars are now found to be younger than the age of the Universe. It is a prime example of how science progresses, iterating towards a solution with better and better agreement between theory and observation. Current age estimates of the Universe are 13.7 Gyr, whereas the age of the oldest-known stars is about 13.2 Gyr.

Einstein's Theory of General Relativity and the Λ Cold Dark Matter Model

Einstein's General Theory of Relativity is the description of how energy and mass serve to distort space-time in such a way as to explain the gravitational interactions between objects. In the limit of weak gravitational fields it reduces to Newtonian gravity, Newton's Laws of Motion, and Einstein's Special Theory of Relativity. It has been used to describe gravitational lenses, black holes, neutrons stars, and other such exotic objects. Most importantly, Georges Lemaître determined in the 1920s that one solution to the Einstein field equations (as the governing equations of general relativity are known) could describe the Universe as a whole, treating the Universe more or less as a space-time continuum that contains mass and energy acting upon itself, and that the Universe could expand or contract with time. Depending upon the density of material within the Universe, the Universe will either a) continue to expand, but at a decelerating rate, eventually reversing, and collapsing back on itself in the future (a Big Crunch!), b) or continue expanding forever. Basically it amounts to whether the Universe can "escape" from itself, or not, in the same way that an object, given an initial boost, either escapes from the Earth's gravity, or falls back to the surface; the answer depends upon the initial boost to the object, and on the mass of the Earth. The Big Bang provided the initial boost, and the self-gravity of the Universe provides the retarding mechanism against expansion. This was the view until the late 1990s and early 2000s when suddenly observations of very distant supernovae determined that on very large scale that the expansion of the universe is *accelerating*! Some sort of pressure within the universe as a whole is overcoming gravity, causing the rate of expansion to accelerate. In our object-leaving-the-Earth analogy, it's the equivalent of strapping a rocket to the object to help it out! We call the source of this pressure *dark energy*, *"dark"* because the only way we've detected it is through this measured acceleration. It is *not* normal matter as we know it, which we would otherwise be able to see.

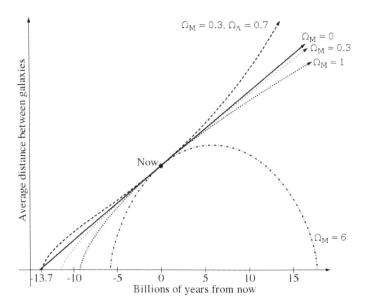

Illustration 13: Various fates of the Universe depending upon the density of matter in the Universe (M). The top-most curve is the model that current observations support. Until the discovery of dark energy in the late 1990s, the other scenarios were thought to be possible, but observations could not distinguish between the various cases.

The current most-accepted model used to describe the universe is the *Λ Cold Dark Matter* (ΛCDM) model, which describes the various components that go into the total energy budget of the universe. They are approximately as follows:

1) 74% - dark energy
2) 22% - dark matter (more on this later!)
3) 4% - ordinary matter (gas, stars, all the stuff we can touch!)

The Λ (upper-case greek letter *lambda*) refers to the *cosmological constant* in the *Friedmann equations* that are the solution to the *Einstein Field Equations*, and is related to the fraction of dark energy within the Universe (so #1, above). This is the part that got suddenly added into the equation in the late 1990s – prior to this point we thought Λ was zero! This was a paradigm shift in the way astronomers viewed the universe, and serves as a cautionary tale to all scientists: no matter what you might think the universe might be like, new observations/experiments that contradict that view always take precedent, and you will need to reevaluate.

Reevaluate, we have, but we're really not sure what dark energy is, just that observations tell us that it exists, so stay tuned for future developments!

Entry #2 above, *dark matter*, is another problem area in astrophysics.

Dark matter, simply put, is matter that we think exists, but cannot directly detect, it neither absorbs nor emits electromagnetic radiation, we only observe it through the gravitational effects it has on the surrounding material that we *can* see. Nor can it be ordinary matter, otherwise (once again) we should be able to detect it directly, rather than just through gravitational effects. Dark matter seems to affect the gravitational field on the scale of galaxies and larger. The most-famous example is that of the rotation speed of galaxies. In disk galaxies we observe stars and gas orbiting about the center of the galaxy at such a rate that we would expect it fly apart if the gravitational field is attributed to only the material we can see! Clearly, there must be more mass than we can see holding things together. Similarly, the velocities for individual galactic members of galaxy clusters are much too high for the cluster to be held together by the amount of regular matter we do see.

The most likely candidate for dark matter is some sort of *cold dark matter*, *cold* meaning the material is not moving close to the speed of light, i.e. is non-relativistic. We do not know the composition of cold dark matter, only that certain types of objects appear to be ruled out. Originally, MAssive Compact Halo Objects (MaCHOs) were one candidate, which are relatively massive objects like black holes, neutron stars, brown or white dwarfs, or "free" planets (objects like Jupiter not gravitationally bound to any star) that had cooled to indetectability, only interacting gravitationally. However, searches for gravitational lensing events caused by such objects (microlensing events), such as the *Optical Gravitational Lensing Experiment (OGLE)*, have shown that events are not numerous enough for such objects to make up a significant portion of dark matter.[5]

---

5   MaCHOs are actually an excellent example of science at work – a testable hypothesis was made that they were a significant source of dark matter. Appropriate observations were made that turned up conclusive negative results – MaCHOs could only make up a very small portion of the missing dark matter, so there must be another dark-matter source.

Currently, the most likely candidate for dark matter is some sort of Weakly Interacting Massive Particle (WIMP – you might be sensing a pattern here with acronymns!), as in some sort of fundamental particle that we don't know about that only interacts weakly with other matter through gravity, not through one of the other known fundamental forces (the weak, strong, and electromagnetic forces). Currently, no known particle fulfills the required properties for dark matter, and it remains to be seen if any candidate will be discovered. Regardless, if any discovery is made, it will mean significant revision of the textbooks!

The Cosmological Principle

The *Cosmological Principle* (CP) is one of the fundamental assumptions upon which we assume the Universe operates. The CP assumes that everywhere in the Universe the laws of physics are identical, and on large scales, the Universe is homogeneous, and isotropic. That is, no matter the direction you look, nor your position in the Universe, any observations that you make should not be influenced by either. On small scales your view may be blocked in certain directions (e.g. we can't see through to the other side of the Galactic core because of all the obscuring material), but step back far enough, and the Universe should look the same in all directions. A consequence is that there is no particular part of the Universe that is special.

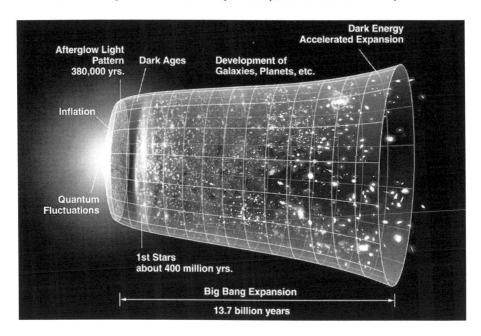

The Big Bang and the early history of the Universe.

We've already alluded to the Big Bang, and suggested that this was the very hot, dense beginning of the Universe, from which the Universe expanded, cooled, and evolved to what we have today. So what was that origin like, and what was the early history of the Universe? What is the observational evidence to back this up?

The history of the Universe is best looked at on a logarithmic time scale, as in the beginning the Universe is expanding and cooling in such a manner.

Scientists have been able to make educated guesses on the state of the Universe back to 10^{-43}s after the Big Bang. Prior to this point the Universe is so small and dense that the laws of physics as we know them, particularly quantum mechanics and Einstein's GR, begin to interact in ways that we do not understand, and so we cannot speculate on what came before this.

As the Universe expanded from this point, we go through many important eras, but in the very beginning normal matter as we understand it (protons, neutrons) did not exist, the Universe was too hot for them to exist, for if such a particle was formed, it was quickly destroyed by collisions with other particles or photons.

Here is a brief list of when various important events occurred.

Grand Unified epoch 10^{-43} to 10^{-36}s - all known forces are unified, i.e. Indistinguishable and acting the same. This includes gravity, the strong nuclear force and the electroweak force.

Inflationary epoch – 10^{-32}s – a period of exponential expansion caused by a phase transition in the Universe. The exact nature of is era is unknown, its former existence not certain, however without it certain problems with the Universe cannot otherwise be explained. A prime example is that the observable Universe is otherwise too uniform without inflation!

Protons, neutrons stabilized – 10^{-6} to 1s – the Universe cools sufficiently to allow neutrons and protons to form without otherwise being destroyed.

Electrons stabilized – 1 to 10s – now it's the turn of the electrons! However, the Universe remains in a plasma state, i.e. despite the existence of both protons and electrons the Universe remains too energetic for neutral atoms to exist.

Nucleosynthesis – 2 to 20 minutes – protons are fused into heavier atomic species, mostly ^{4}He, but also trace amounts of ^{2}H, ^{3}H, and ^{7}Li. The Universe expanded too fast, and the energetics were not favorable for the formation of heavier species. Most of the free protons survived, leaving approximately 75% hydrogen, but 25% by mass is incorporated into ^{4}He, and 0.01% to ^{2}H. These are approximately the abundances we observe today, modified by stellar-evolution processes over time. The mass fractions impose important limitations on any model of the Universe, and the detailed physics that go into making up those models.

Illustration 14: Cosmic Microwave Background (CMB) as measured by the Wilkinson Microwave Anisotropy Probe (WMAP).

Recombination and the Cosmic Microwave Background – 400,000 years - around this time the Universe has expanded and cooled sufficiently to allow neutral hydrogen and helium to form without immediately being re-ionized by a collision with another particle or photon. The temperature is around 3000 K. Suddenly, the universe essentially becomes transparent to most radiation! And, indeed, most of the radiation from that era still exists in the form of the Cosmic Microwave Background (CMB), discovered by accident Penzias and Wilson 1964. Via the expansion of the Universe, that radiation has been "stretched" to much longer wavelengths than at the time it was produced. The peak wavelength at the moment is 1.87mm (microwaves), corresponding to a black-body spectrum of 2.725 K, and is the current temperature of empty space! The CMB is one of the cornerstone pieces of evidence in support of a Big Bang origin for the Universe. Photons produced prior to this era would have interacted with matter, and thus no longer exist. This is the earliest direct glimpse of the Universe that we have. Anisotropies in the field (differences between one direction and the next) have implications for the structure of the universe at the time, and thus how such things as galaxies and galaxy clusters will eventually develop.

Smoothness of the Universe – at earlier epochs matter and radiation pressures were interacting in such a way as to smooth out inhomogeneities that might form. As time goes on, matter interactions (including dark matter) start to dominate over radiation interactions, and inhomogeneities start to form, leading to the "seeds" that will eventually collapse to form the earliest-known objects.

The very first stars – as previously mentioned when talking about the Milky Way Galaxy, the first stars seemed to have formed before the Galaxy did. Moreover, the first generation of stars could not have contained any elements not produced in the Big Bang. Stars bereft of metals are known as Population III (Pop. III) stars. The evidence suggests that these must have been very large stars (maybe 100 $M_\odot$+), with a formation mechanism different than what we observe today for less massive stars (an ongoing research topic). Indeed, no small-mass stars (less than 0.5 $M_\odot$, say) seem to have been produced, otherwise these stars would still exist, and we would be able to detect them! Remember, the less massive the star, the longer its lifetime, with massive stars living very short periods of time, indeed! *So far, every star that has been analyzed in detail has metals in them, and is thus (at least) a second-generation star.* This initial period of star formation was probably quite intense, laying the groundwork for future generations of stars by producing the first heavy elements, scattering the elements to the ISM via radiation-driven winds and supernova explosions, and producing the first black holes. The black holes are important, see below.

Illustration 15: X-Ray image of the quasar PKS 1127-145, distance about 300 Mpc.

Quasars and early galaxies – light has been detected from galaxies that is 13.2 Gyr old, or produced about 500 million years after the Big Bang. These tend to be amorphous blobs of bright light, areas of intense star formation, and gas interaction, not like the structured galaxies we see in more modern times. *Quasars* are often associated with such objects. Quasars (a slang term for quasi-stellar object, as they can look like stars in low-resolution images) are a class of *active galactic nucleus* (AGN) that are intensely bright, powered by an accretion disk falling on to a massive black hole at the center of the host galaxy. How do the black hole form initially, given that no such objects are thought to have been produced by the Big Bang? Unknown, but one guess is that smaller black holes from the first generation of massive stars joined together to form a central black hole, which then dominated the local region, accreting dust and gas, giving off the light we eventually see 13+ Gyr later! Eventually, these black holes became the centers of large galaxies, continuing to grow with time.

Active Galactic Nuclei

We've already encountered the Super Massive Black Hole (SMBH) residing at the center of the Milky Way Galaxy. There is reason to believe that perhaps all massive galaxies contain a SMBH, and that occasionally material is swept into the center of the galaxy, forming an accretion disk around the black hole. The material produces lots of radiation as it loses enough energy and angular momentum to falling into the black hole. We detect this radiation, and these objects are collectively known as active galactic nuclei (AGNs). The amount of in-falling material, the size of the black hole, and the orientation of the observer with respect to the spin axis of the accretion disk influence what we see. Indeed, the disparate phenomenology associated with AGNs required much research before it was determined a basic mechanism was at work in all cases, that of material accreting onto a black hole.

Quasars are the most powerful example of an AGN, and are mainly associated with newly-forming galaxies in the early Universe, usually outshining the entire nascent galaxy. Low-Ionization Nuclear Emission-line Regions (LINERs) are perhaps the least-energetic example, only detectable through emission lines observed from the core of the host galaxy. Jets of material originating from the core of the AGN may also be observed, as in the case of the galaxy M87 in the Virgo cluster. Emission can be observed all across the electromagnetic spectrum, from radio waves to gamma rays.

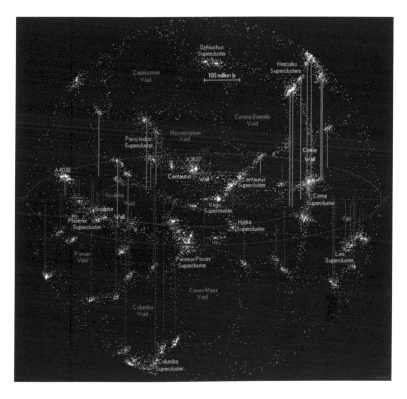

Illustration 16: Current structure of the universe showing clusters, and superclusters of galaxies. The Local Group is at the origin, and is too small to be seen!

Modern Structure of the Universe

Eventually, through galaxy mergers, and accumulation of material into filamentary structures, we end up with the modern-day Universe, with massive galaxies, galaxy clusters, and superclusters of galaxies. We can only guess how it will eventually turn out, but we can expect further galaxy mergers, such as the Andromeda and Milky Way galaxies merger already mentioned. In the very long term we might expect dust and gas to eventually be locked up in less-massive stars, black holes, and stellar remnants (white dwarfs, neutron stars), with the Universe not so bright a place any more.

To date, this is how astronomers and astrophysics currently explain the Universe, and our place in it. There are many questions still to be answered, but our understanding of the Universe has been transformed since Galileo first pointed a telescope at the night sky, and Kepler and Newton figured out how and why the planets move the way they do. Stay tuned for more revelations as telescopes improve, more discoveries made, and the theorists struggle to keep up with it all!

5054340R00023

Printed in Great Britain
by Amazon.co.uk, Ltd.,
Marston Gate.